60-Minute Career Path Deep Dive

CYBERSECURITY®

Books by C. L. Freeman

Master Distance Learning – 2016
Master the Interview – 2018

60-Minute Career Path Deep Dive
CYBERSECURITY©

C. L. FREEMAN

A revised edition
with input from additional industry recognized
Cybersecurity practitioners

C⁴ Series

Dedication

I dedicate this book to our Lord Jesus Christ, several United States Navy Sailors and employees of The Boeing Company who helped influence my evolving strategy to build a solid foundation for a 35+ year career in Cybersecurity.

Forward

- _**Cybersecurity (CyberSec) is One of the Hottest Career Paths in 2021 and Has Been Since 2005**_

The number of unfilled job openings continues to grow. The only factor that appears to be tempering its accelerating rate of growth is the willingness of Federal/state agencies, academia and private industry to transfer account-ability to protect sensitive client, enterprise or employee information to Cloud Service Providers (CSPs) like Microsoft and Amazon Web Services (AWS).

Like anything in demand, people naturally gravitate to it over time because it has a high probability of a tangible, realized Return on Investment (ROI). High school students, college juniors and those mid-career with no experience in Information Technology (IT) or Cybersecurity are seriously considering re-en-gineering their career strategy to enable them to break-into the Cybersecurity career path.

- _**People Want to Know More About the Realities of a Career in Cybersecurity**_

Specifically, they want to know a) what steps they need to take to get an entry level position, b) what strategic moves will be critical mid-career for a CyberSec professional and c) what steps must be taken the last ten years of a career in this domain to ensure they are highly valued by their current or potential employers.

- _**People Want to Hear from Senior Leadership Who Have Sound Advice for Those Already Working in or Seeking a Career in the Cyber-Sec Domain**_

This career deep dive is an attempt to communicate answers to these questions and more. The answers are provided by CyberSec professionals working in academia, defense and space, commercial industry and the U.S. Military. Some have served as Chief Information Security Officers (CISOs), Chapter Presidents of IT Security professional organizations, Senior U.S. Military Cy-berSec professionals, Directors of Information Security, Corporate Information Assurance Officers (CIAOs), Directors of Export Compliance, System / Network Security Architects, System Security Engineers (SSEs), etc.

Thanks to the multiple current and former employees of the following Federal/state agencies and commercial firms for providing input for this book:

- RAND Corporation
- United States Navy
- Maui High Performance Computing Center (MHPCC)
- Microsoft
- GoPro
- United States Army
- Defense Counterintelligence and Security Service (DCSA)
- The Boeing Company
- University of Arkansas
- The College Board
- Booz Allen Hamilton
- First Financial Northwest Bank

Carl L. Freeman, CISSP-ISSAP

President, (ISC)^2 Orange County Chapter

Contents

-1-

"To retain Cybersecurity talent in your organization, be sure to give employees more money, as appropriate, linked to their performance. Be sure to regularly communicate how they fit in the organization, provide recurring personalized career coaching and try to give them a glimpse what their future in the organization will be a few years ahead. Also, don't discourage someone who is considering leaving your organization if that move has strategic value in context of their short and longer term career aspirations. Don't forget to periodically communicate their value to you and their contribution to the success of the organizations Cybersecurity program. If they are a top performers, let them know frequently and honestly."- **Todd Barnum; Chief Information Security Officer (CISO); GoPro**

"Demand for Cybersecurity talent is at an all time high in 2021. Management must meet with their Cybersecurity staff face-to-face, bi-weekly if at all possible to proactively determine employee satisfaction with a) their work environment, b) opportunities for training and c) management's commitment to assign them tasks that enables sustained development of new skills that will improve their value to the company / likelihood of advancement in the organization. If this strategy is not applied at a consistent battle-rhythm, management shouldn't worry about if Cybersecurity employees may leave the organization, they should simply try to predict when they will depart" - **Carl L. Freeman, former Director, Security; Maui Space Surveillance Site (MSSS)**

What top three actions can leadership take quickly to reduce the risk of a high rate of employee turnover in the Cybersecurity career path?

1. **Assign a Mentor with Superior Skills:** Employees in this domain common-ly view working with a subject matter expert with superior experience and

1

skills as critical to their long term career development. Assigning these employees to a senior person on the team day one and well beyond with likely influence their longevity with the organization.

2. **Understand Market Rate Salary and Sign On Bonus Realities for Staff with Relevant Skills, Formal Education & IT Security Relevant Certifications:** Most medium to large companies do a good job of gathering valid metrics concerning competitive salary rates for Cybersecurity staff or leadership, however, small companies do not. If Human Resources and company leaders of Cybersecurity professionals on their team don't proactively gather this data annually, the significant time and financial investment in career / skill development for their Cybersecurity staff may end up benefitting a competitor.

3. **Secure a Respectable Budget for Technical Training and Conferences for Your Cybersecurity Staff:** Cybersecurity staff realize the value of their training in the domain has a shelf-life. Like operating systems Microsoft no longer supports, Cybersecurity staff must be offered the opportunity to refresh and learn new relevant skills or their value to the organization and more importantly, their viability / credibility as a professional to peers in the domain will suffer. Cybersecurity professionals don't have an option. They must work for employers who understand the vital importance and pro-actively fund ongoing technical training and attendance at relevant industry recognized conferences, workshops or IT security certification boot-camps.

"If at all possible, try to stay with any organization you work with for a minimum of two years. Leaving before 24 months can leave the impression by those who review your resume you are job-hopping to secure a better salary at your next company or you don't stay long with any employer because you have a challenge you won't likely communicate on your resume or in an interview"
- Carl L. Freeman, former Chief Information Security Officer/ Director, Information Security; RAND Corporation

-2-

"Ensure your resume notes execution of as many tasks as possible, in current or past positions, that link directly to tasks noted in the employer's written job description. You must resist the temptation to claim experience, security relevant IT certifications or education you don't have. If employers determine this during a face to face interview, the long term damage with this potential employer will likely be permanent" – **Alan Greenberg, former Chief Information Security Officer (CISO); University of Arkansas**

"Ensure you don't consume too much real estate in your resume to simply list the administrative, business or technical tasks of your current or past security/non-security relevant positions. Be sure to note two to three 'contributions of significance' during your tenure for all positions in your resume that enabled the businesses you supported to execute a key process better, faster or cheaper. In other words, highlight those victories achieved that your employers will benefit from for years to come" – **Carl L. Freeman, former Corporate Information Assurance Officer (CIAO); Booz Allen Hamilton**

What key content is of highest value in a resume for a Cybersecurity position?

Three elements of content in candidate resumes are of particular value for all CyberSec positions, regardless if they are staff, management or senior executive level opportunities. Noting all formal education and Information Technology (IT) or security relevant certifications such as CompTIA Security+ or the (ISC)² awarded CISSP or CCSP certifications are very important. Equally important is a brief summary of the tactical level responsibilities, in addition to a brief description of 'Achievements of Significance' for all positions noted in the resume.

Most resumes are simply a series of verbose paragraphs or bullets simply highlighting the daily or weekly duties and accountabilities for positions the candidate has held over the relevant number of years.

Ensure you note two or more examples of real technical, business or administrative processes you singlehandedly created, streamlined or improved that resulted in cost or cycle-time reduction, team's you led at the company or agency that resulted in an award directly to you or the team from senior management.

Organization Human Resources staff will likely spend 30 seconds reviewing your resume to determine if they should forward the resume on to the Hiring manager. Some HR representatives don't review your resume at all because they rely on an algorithm baked into the companies resume web tool of choice to determine candidates who were smart enough to add at least 5-10 key words from the job description posted on the internet to avoid being eliminated from contention well before HR reviews the typically large pool of semi-qualified candidates.

Hiring managers will likely spend less then 2 minutes per resume they review to decide if your resume will make it into the top 3-5 'Downselect List' that will get deeper level attention as management decides who will be invited to a virtual interview and eventually, a face to face exchange so they can determine if you likely shower daily.

Since the hiring manager is going to spend so little time reviewing your resume, it's likely best to note your education and security relevant certification(s) summary on page one of the resume and strive to honestly note those tasks you execute currently, or did in your last 2+ positions that are in the job description provided by your potential employer.

"Closely monitor the shelf life of education you note in your resume. If you haven't noted any additional education in your resume for 5+ years, some hiring managers will conclude you won't proactively take the initiative to understand the organic security risks that will apply to technologies the organization will need to embrace in the future to maintain competitiveness"
— **Carl L. Freeman, former Corporate Information Assurance Officer (CIAO); Booz Allen Hamilton**

-3-

"What matters most to a Cybersecurity professional is not who you know, but who knows you. Teaching, especially online, is one of the best ways to establish your professional standing in this rapidly growing community. Teaching Cybersecurity online gives you the best way to reach the most people who could benefit from your cyber security knowledge and guidance." – **Bill Nelson, former Computing Forensics Investigator; The Boeing Company**

"If you have the opportunity to teach Cybersecurity at any academic institution, you should take advantage of it. You should pro-actively integrate the goal of teaching related courses online or off campus into your long term career management strategy" – **Carl L. Freeman, former Director, Information Security; Maui High Performance Computing Center (MHPCC)**

Should you consider teaching Cybersecurity online or on-campus at the community college or university level?

Teaching curriculum directly related to Information Technology (IT) or IT in the context of Cybersecurity will be of particular value to your effectiveness as a CyberSec practitioner. The benefits are many. Here are a few:

- Your perceived value to current or future employers will be enhanced, which will likely impact your starting salary, annual salary increases or bonuses when compared with peers

- Teaching CyberSec also has value in helping you exercise skills real-time you'll eventually require when you must brief your company Board of Directors or Chief Information Officer (CIO) concerning relevant threats to the corporate network, current technologies deployed or budget required to enable fielding of technical countermeasures to mitigate relevant risks to company networks or data

- Teaching CyberSec also serves as a motivation to proactively refresh your understanding of evolving technologies vendors offer to mitigate relevant risk

- Teaching on campus or online CyberSec curriculum offers a buffer of revenue if you temporarily lose your job, retire or simply desire a part-time ongoing employment opportunity to avoid boredom.

- You should also consider teaching IT security certification preparation courses as well. Courses that prepare candidates to sit for CompTIA Security+, (IS2)2 CISSP or CCSP exams, for example

- You should also consider taking advantage of opportunities to teach vendor developed content for students preparing to take industry recognized IT Security certification exams such as CISSP, ISSAP, CAP, CCSP, CISM or CompTIA Security+

"If you hope to move into management in the future, strive to fully understand what your managers need from you to keep their senior leadership happy.

*Also, seek a mentor, other than your manager in the organization and also ask your manager to coach you concerning how to formulate strategy, document performance metrics or business / budget related deliverables they are accountable to deliver to their leadership on a recurring basis" – **Carl L. Freeman, former Director, Information Security; Maui High Performance Computing Center (MHPCC)***

"The quickest way to form a foundation that will enable you to compete for an entry level position in Cybersecurity: 1) Take a few community college courses in Cybersecurity topics at a local community college, 2) Study for and pass the CompTIA Security+ certification and 3) Find a mentor at work or outside your present workspace who will spend time with you during or after hours to learn practical application of what you're learning as you pursue the other two recommendations" - **Mike Waters, former Director, Global Information Security; Booz Allen Hamilton**

"You've got to accept the reality you'll need to spend a significant amount of your personal time, at the keyboard, learning how to use leading edge Cybersecurity applications and tools. This will enable you to effectively communicate a deep-depth understanding of how they work under the hood if you get the opportunity to sit for an interview for a position in this domain" - **Carl L. Freeman, former Chief Security Architect – FIA Program, Ground Segment; Boeing Defense & Space Group**

I'm a mid-career, degreed financial professional with a serious interest in Information Technology (IT) and a desire to pursue a Cybersecurity career change. What are the top three things I could do if I want to transition into Cybersecurity quickly?

1. **Find a Way to Learn and Apply Skills In the Short Versus Long Term:** Consider joining the U.S. Navy or Army Reserves in a Cyberwarfare role. Enlisted or Officer career paths will be of short term civilian career benefit because you'll likely be required to complete relevant Cyber Training delivered by NSA or other U.S. Intelligence Community (IC) agencies. You'll likely also be eligible for a high-level U.S.

Government security clearance which will enable you to pursue a wider range of employment opportunities with U.S. Government agencies or Cleared Defense Contractors (CDCs) such as Boeing, Booz Allen Hamilton, Northrop Grumman or RAND Corporation.

2. **Quickly Add Relevant & Reputable Education to Your Resume:** Consider completing a few Information Technology (IT) Security certificate programs from a nationally respected reputable institutions. This will differentiate you from your competitors when you apply for and interview for cybersecurity positions. For example, Stanford University offers an online Advanced Computer Security Certificate you can complete in a few months.

3. **Find a Technical Mentor and Invest Time with Them:** Consider finding a mid-career professional in the cybersecurity domain who can serve as a mentor to guide your self study or professional training choices. This will likely result in reducing the risk of applying your limited time and money to strategies that will not result in skills, training or IT security certifications, such as CompTIA Security+, employers won't value in a resume. Building a home computing network and asking a well skilled system administrator to help you harden and manage the configuration for an hourly rate is also another method to streamline your hands-on training strategy.

"Confidence in your technical knowhow or abilities can be a blessing or a curse.

Many with deep technical ability fail to acknowledge the line between confidence and arrogance is often hard for them to see, but their peers, clients and leadership see it clearly. Too much confidence can stifle a bright career." - **Carl L. Freeman, former Chief Security Architect – FIA Program, Ground Segment; Boeing Defense & Space Group**

-5-

"The top Cybersecurity employment opportunity most don't know about or rarely take advantage of that will open a wide range of opportunities with just a few years service is a U.S. Military Reservist. You can get leading edge, industry respected Cybersecurity training, practical experience during drill weekends and in many positions a TOP SECRET security clearance" - **Anonymous**

"Employment opportunities in Cybersecurity are very diverse. Not all require technical depth to have a successful career. Corporate or academic institution Information Security Policy, Standards & Procedure development is an example of a high value skill that doesn't require heavy technical training, a computer science degree or mastery of leading edge security relevant COTS applications"- **Carl L. Freeman; President, (ISC)2 Orange County Chapter**

What are the top forty Cybersecurity employment opportunities in 2021?

1. Splunk application administrator
2. Manager, Information Security
3. Systems Security Engineer (SSE)
4. Network Security Architect
5. Assessment & Authorization (A&A) Professional - CNSSI 1253
6. Chief Information Security Officer (CISO)
7. Information Technology (IT) Security Policy Analyst
8. HP ArcSight ESM application administrator
9. Secure Mobile Applications Developer
10. Cyber Warfare Analyst (U.S. Department of Defense)
11. Director, Information Security
12. Advanced Persistent Threat (APT) Detection & Analysis
13. Information Technology (IT) Security Consultant
14. Cybersecurity College Professor / Instructor

15. IT Security Certification Boot Camp Instructor
16. Chief Security Architect
17. Network Penetration Test Engineer
18. Secure Code Analyst / Test
19. Cloud Security Analyst / Engineer
20. Cyber Operations Center (CSOC) Lead / Manager
21. Network Forensics Investigator
22. Account / Access Administrator
23. Security Help Desk Analyst
24. Cyber Metrics Analyst
25. Cryptographer
26. Cyber Insurance Policy administrator
27. Ethical Hacker
28. Malware Analyst
29. Governance, Compliance & Risk Manager
30. Red Team Cyber Analyst
31. Blockchain Developer
32. Chief Security Officer (CSO)
33. Cybercrime Investigator
34. Cybersecurity Hardware Engineer
35. Cybersecurity Attorney
36. Disaster Recovery Analyst
37. Industrial Internet of Things (IoT) Security Specialist
38. PKI analyst
39. Vulnerability Assessor
40. Incident Response Analyst/Manager

"A Bachelors/Master's degree or certificate from an institution with a poor reputation can result in little value in the long run. An academic institutions poor reputation can be earned in several different ways.

The most prevalent is low quality professors that don't teach their students 'systems thinking' or how to conduct quality research. Some have a gained a reputation as a 'Diploma Mill' and that reputation will be hard to shake for years to come.

Be careful to ensure you attend institutions that have a solid reputation. It will help you effectively compete with candidates who apply for the same opportunities you do in the future."
- Carl L. Freeman; President, (ISC)² Orange County Chapter

60-Minute Career Path Deep Dive

CYBERSECURITY©

-6-

*"Research skills are often an overlooked skill in the cyber security industry. The ability to quickly research, understand and interpret banking regulations, standards and policies is crucial to advancing a career in CyberSec at all levels. A CyberSec professional must continue to improve their research skills, know where to find information and be able to provide the correct information to management. There are many different roles within Cyber Security such as technical, administrative, regulatory, audit and managerial to name a few. You may find yourself switching roles several times throughout your career by your own accord or due to organization needs and requirements. Having knowledge of the disciplines in other Cyber Security roles will prepare you for changes in your career path and help to minimize the pressure and anxiety of a new role."- **Don Boelling, Vice President – Security Officer; First Financial Northwest Bank**

*"A CyberSec professional must proactively define what their resume will say two years ahead in three primary areas – Skills, Industry Recognized Information Technology (IT) Security Certifications and formal IT education. The skills and education roadmap will change over time. At the senior level, business and political skill will likely be of higher importance"- **Carl L. Freeman, former Director, Security; Maui Space Surveillance Site (MSSS)**

What key skills, certification(s) or formal education contributes to a successful Cybersecurity career in the banking industry?

Skills: Foundational skills in managing a client or server operating system are of particular value. Knowing how to configure and interpret output generated from the evolving portfolio of COTS products that can be fielded on all network appliances, servers and endpoints to detect the insider threat or an attack from an external entity will also be of high value. A demonstrated commitment, visible in language of the resume to 'Lifelong Learning' is also critical to a long-term successful career as an Information Technology (IT) security analyst, manager or senior executive in the banking industry.

Certification(s): At entry level, CompTIA Security+ is a valuable Information Technology (IT) security certification. Mid-career, the CISSP certification will also be valuable in the banking industry. It's also probable more banks and the U.S. Federal reserve will gain higher confidence in Cloud Service Providers (CSPs) as a potential source to provide IT services at a significantly lower cost. An (ISC)2 CCSP certification may prove to be of value as well if leadership intends to sunset their on-premises IT and migrate their operations to Amazon Web Services, Microsoft Azure or other CSP solution.

Formal Education: A bachelor or graduate level degree in Information Systems with a concentration in Information Security (INFOSEC) will be of value. A Computer Science degree with a CyberSec minor is also of high value in the banking industry. Certificate programs are also of benefit if they are awarded from respected institutions and the series of Cybersecurity courses contributes to improving the value and relevant skills of the employee.

"A Cybersecurity professional needs to acknowledge early in their career well developed writing skills will be critical in the early, mid and latter phase of a 40+ year career.

Strive to learn and apply changing best practices concerning development of written content for senior level presentations, research summaries, cost/benefit analysis summaries, e-mail to leadership and external clients, etc."- **Carl L. Freeman, former Director, Security; Maui High Performance Computing Center (MHPCC)**

7

"For disk forensics, Autopsy Sleuth Kit for Windows. This is a freeware tool and is routinely updated with significant improvements and features. Other freeware and demo-ware tools are hexadecimal editors such as HxD Hex Editor (freeware) and WinHex (demo-ware, 30-day free use before warning message appears)" - **Bill Nelson, former Forensics Computing Investigator; The Boeing Company**

"The forensics toolbox will change over time, so it's important to read job descriptions posted on the internet and participate in IT Security professional organizations so you can plan your training strategy in a manner that is most valuable to your current and future employers" - **Carl L. Freeman, former Chief Information Security Officer (CISO); RAND Corporation**

What moderate cost tools can high school or college students obtain and learn to use that would serve as a good foundation for understanding the fundamental features found in network or computing forensics tools used by the FBI, State police or corporations with an Information Technology (IT) forensics investigations team?

The answer to this question is quite different if you are talking about network versus server / client operating system platform or mobile phone forensic investigations.

If you want to learn about forensic network investigations, WireShark is a good tool to start with. YouTube has series of videos that address how to exploit key features of the product and user documentation on their website is above average.

Autopsy Sleuth Kit is another tool one can learn apply at home or school for server/workstation drive forensics. YouTube has several tutorials for beginners or those who seek more advanced skills with the product. The Sleuth Kit is a collection of command line tools and a C library that allows you to analyze disk images and recover files. It is used behind the scenes in Autopsy and many other open source and commercial forensics tools.

WinHex is in its core a universal hexadecimal editor, particularly helpful in the realm of computer forensics, data recovery, low-level data processing, and IT security. An advanced tool for everyday and emergency use: inspect and edit all kinds of files, recover deleted files or lost data from hard drives with corrupt file systems or from digital camera cards. Features depend on the license type.

"Technical skill and implementation of a pro-active strategy to maintain a relevant portfolio of skills is critical to a Cybersecurity professional's success.

However, 'Soft' skills are of equal value and must be cultivated to ensure their heath as you navigate through the different phases of a long career.

An over-confident, arrogant Cyber Technologist will likely succeed early in their career. However, their lack of experience encouraging others to succeed, motivating a team, mentoring a junior employee, exhibiting a commitment to ethical conduct and leading by example will result in career stagnation that eventually leads to the employee or organization parting ways."
- Carl L. Freeman, former Chief Information Security Officer (CISO); RAND Corporation

8

"You MUST have the mindset of a CSO/CISO. Do whatever it takes to develop your IQ, PQ (positivity quotient) and EQ (emotional intelligence quotient). Job descriptions in requisitions are generally terrible. Don't rely on this. Be VISIBLE via LinkedIn and Twitter. There are specific barriers that affect women in this industry differently than men. INVEST IN YOURSELF by hiring a coach to help you accelerate: https://createyourleadingedge.com, for example."
- Karen F. Worstell, former Chief Information Security Officer (CISO); Microsoft Corporation

"If possible, find one or more CISO mentors who can offer you constructive criticism about your ongoing efforts to eventually qualify to serve as a CISO. Also, read constantly about Cybersecurity to increase the width and depth of your domain vision"
– Carl L. Freeman, former Systems Security Engineer (SSE) – French AWACS; Boeing Defense & Space Group

I hope to be a Chief Information Security Officer (CISO) someday and have 4 years experience in the domain. Assuming I have a Masters degree in Cybersecurity from a reputable university and an (ISC)² CCSP certification, what strategies should I execute in the next 5 years to prepare myself to compete for a CISO position?

Initially, you need to accept the fact it will likely take 15 years of effort to develop the foundation necessary to apply for a CISO position. If you want to be the CISO of JPMorgan Chase, RAND Corporation or The Boeing Company, your resume will require different content if you hope to sit for an interview. A head hunter firm is often hired to find the best

CYBERSECURITY 2021

candidates in the relevant industry and they'll likely contact several but nominate few. A few high-level strategies will be critical, regardless of the companies you consider. For example:

- **Seek Opportunity for Promotion Every Three Years** – You are going to need to have a track record of success at staff and lower-level management positions in the domain. More importantly, 3+ years at the senior management level.

- **Accept That You May Have to Leave Your Company** – If your promotions are not leading to an appointment as a manager or senior manager, you need to embrace the reality you need to move on to another company.

- **Maintain a Consistent Commitment to Lifelong Learning** – Refresh your resume with new, relevant certificates, IT security certifications, etc. An (ISC)^2 CCSP certification and 'Advanced Computer Security Certificate' from Stanford University will separate you from the crowd of applicants. Also, review CISO job descriptions regularly to get an idea of the evolving technical, business and management skills required to apply.

"Cybersecurity professionals, especially those who love to spend 99% of their time at the keyboard versus dealing with people, often overlook the importance of their professional appearance. For example, they don't see value in ever wearing 'business casual' and when appropriate, a suit and tie. They may wear AC/DC or 50 Cent tee-shirts and jeans to work and sometimes don't trim their nose hair well before someone asks them to seriously consider it.

*A mentor with intestinal fortitude may take them aside and make it clear how this choice shapes their brand and more importantly, their perceived value to the organization." – **Carl L. Freeman, former Systems Security Engineer (SSE) – French AWACS; Boeing Defense & Space Group***

-9-

Where can the Cybersecurity professional learn offensive Cyber Warfare techniques?

All nation states, except the poorest of nations on the planet are vulnerable to Cyber attack by their enemies and as we've seen a few times in the media over the last 30 years, allies as well. Nation state sponsored attacks against a country's communications or power grid, penetration of defense & aerospace corporation intranets that enable data file exfiltration of sensitive defense article design information represent a few of many relevant examples that apply in 2021.

Companies and Government agencies field defensive technologies and train staff to operate a wide variety of hardware and software to counter this treat. Nation states also have an obligation to develop offensive cyber tools as well and train a team of competent staff to use this arsenal, when justified in the interest of national security. Cyber attacks by enemies and allies are considered by most in the Cyber domain an 'Act of War' that justifies delivery of an offensive and destructive response.

Where can a CyberSec professional, living in the United States, learn how to create or use offensive Cyber Warfare applications? The U.S. Army, U.S. Navy and other agencies in the U.S. Government offer employment opportunities as a full time employee or a Reservist that will by default provide professional training and tactical level experience with the evolving tools in the offensive arsenal. Some U.S. defense contractors have been hired by the U.S. Government to create, test and support delivery to the Warfighter as well.

"Cybersecurity professionals cannot simply look-away when an organization or individual has decided to ignore ethical lapses in judgment or implement a strategy that breaks state or Federal law.

You will eventually discover an organization has decided to 1) not pay software vendor license fees for a COTS application, 2) not report a recent compromise of customer Personally Identifiable Information (PII) to the Government or 3) a peer provides you an unauthorized copy of interview questions 'under the table' you'll be asked during an interview for a job you've applied for in the organization.

How you respond when you are made aware of these situations will be critical. If you remain quiet or actively encourage others to modify or delete evidence, your career in the organization or possibly in the career domain as well could come to an abrupt end."- **Carl L. Freeman, former Data Processing Technician, 2nd Class (DP2 AW)); United States Navy**

-10-

*"It's always good practice to consider multiple security frameworks, especially if your enterprise must protect multiple types of sensitive customer, company and employee data via application of specific State or Federal IT security policy (ISO 27K, NIST 800-53r4, CIS 20). Each baseline, when applied, motivates leadership to honestly acknowledge if they are effectively managing the basics, like control of all changes to the enterprise network hardware / software baseline. Most readily admit, following every annual self-audit, they need to do better in this elementary area and acknowledge the reality 'You can't protect what you don't know about"- **Frank Burdette, former Senior Director, Information Security Compliance; The College Board***

*"All organizations should seriously considering gathering and evaluating CIS 20 based compliance metrics annually, even if they are committed to application of another enterprise Information Technology (IT) security requirements baseline, such as NIST 800-53 or ISO/IEC 27001" - **Carl L. Freeman, former Chief Information Security Officer (CISO); RAND Corporation***

What are the SANS **20** Controls and what value is there in knowing the details?

The SANS Institute created and posted the 'SANS 20 Critical Security Controls' on their website for several years. SANS leadership reached an agreement with the Center for Internet Security (CIS) to continue refinement and promote wider application of the controls baseline. Version 7.1 of the CIS Critical Controls baseline has been released to the general public.

The primary utility of the CIS 20 is its value as a high quality security requirements baseline Directors of Information Security or Chief Information Security Officers (CISOs) can apply in their environment to protect on premise or cloud based systems, critical applications or sensitive enterprise, supplier or client data. In other words, it is a requirements baseline similar to other industry and government recognized Information Technology (IT) security baselines like NIST 800-53, revision 5 or ISO/IEC 27001.

Several security leaders in the U.S. Government and industry agree if they choose to apply a different security requirements baseline, analysis of all future releases of the CIS 20 is a best practice. They execute 'Gap Analysis' or rely on other reputable service providers who have applied to effort.

"Ten to fifteen years into a multi-decade career in Cybersecurity, most will hold multiple positions, possibly in multiple companies. Regardless if you stay in the company or organization you begin your career with, it's always a good idea to step back and consider where you did well and where you could have done better in each assignment.

The old saying 'We Learn by Failure' applies in this context. You may have a few 'crash and burn' experiences along the journey. The key is to resist the temptation to only consider what your manager or the organization did wrong that led both parties to that place.

Take the time to document and ponder deeply the full scope of what you did/should have done that contributed to an ending neither party wished for. Take those lessons learned forward to future opportunities" - **Carl L. Freeman, former Chief Information Security Officer (CISO); RAND Corporation**

60-Minute Career Path Deep Dive

CYBERSECURITY©

*"Integrity, unwavering / demonstrated commitment to others versus yourself, consistent delivery on your commitments, on-going pursuit of learning opportunities that benefit your employer and enable achievement of your long term career objective"- **Anonymous***

*"Demonstration of a 'People First' philosophy, consistently, is the most important element of success for any manager. If you simply strive to 'serve' your team, opportunities to succeed will happen. Some simply don't have this trait in their DNA and don't proactively self-measure the appropriate performance criteria that applies"- **Carl L. Freeman, President; (ISC)² Orange County Chapter***

What are the top four skills in 2021 that enable promotion to management in the Cybersecurity career path?

1. **People First Focus:** Some who seek management appointments forget that people, including peers, middle management, internal / external clients and senior organization leadership will determine if they succeed or don't in the long term. You'll need to display a dedication to the success of others first, versus yourself, early in your career if you hope to be selected when interviewing for a first level, senior or executive level position.

2. **Understand the Business and Your Organizations Role in Achievement of Business Goals:** To succeed as a manager, you'll need to understand the basics of business management you can learn in an online business management or finance online certificate from a reputable college or university. More importantly, you'll need to proactively take advantage of all opportunities to learn and if possible apply with mentor input business processes that apply to the company or organization you support

3. **Take the Time to Know What Your Cybersecurity Tools Do Well and What They Don't Do Well:** Dedicate time in your calendar to sit down with your security appliance or utility administrators to understand what your tools can do and if you are getting the full risk reduction value / return on investment for all tools in the toolbox

4. **Proactively and Consistently Shape Your Brand:** A high value personal brand is shaped by never being late to meetings, treating all with dignity and respect, consistently meeting commitments to your peers and internal / external clients, creating high quality products and services, etc.

"For the first six months of any new Cybersecurity assignment, work 12 hours per business, 5 days per week versus the standard 8 hours without requesting compensation.

Utilize the extra time to not only get your daily tasks done that couldn't be completed during the day due to competing priorities, but read organization Cybersecurity policy, document your evolving understanding of success criteria for your job, document what you know well and need to know in the future to add value.

Review organization standards / procedures, study the organization chart to understand data input/output from peer organizations/external entities that play a role in execution of your organizations 'Statement of Work', author your own 'Desktop Procedures' for tasks you find challenging, document your interpretation of 'performance metrics' you can review with your manager at a later date and document / constantly update your strategy to improve your value to the organization and quality of your deliverables.

*This 180 day effort will pay dividends in the longer term. It will play a role in shaping your brand with key stakeholders in the organization and most importantly, you will have a deep understanding of the full scope of performance expectations from all relevant internal clients"- **Carl L. Freeman, President; (ISC)²*** **Orange County Chapter**

-12-

"One of the easiest ways to learn about a Cybersecurity tool is to ask the application administrator to take you under their wing and teach you what goes on under the hood. Video's and textbooks have value, but you can't do better than real time at the keyboard taking the product for several test drives – **Anonymous**

"Junior, mid-level, senior and executive level Cybersecurity practitioners should reserve time every week, outside their normal working hours to actually utilize leading edge security relevant tools if they hope to remain relevant and more importantly competitive in the rapidly changing CyberSec domain" – **Carl L. Freeman, former Chief Information Security Officer (CISO); RAND Corporation**

What are the top three Cybersecurity tools I should teach myself to use or ask a peer to teach me how to apply to my home computer network if I hope to remain competitive or pursue the Cybersecurity career path?

Splunk or Wireshark are good choices. There is also a reduced version of EnCase that can be downloaded for a 30 day trial to learn how to forensically analyze content on a hard or solid state drive. Vendor or subject matter expert developed training resources for all three of these products are available via vendor websites, SkillSoft, Books24x7, Safari Books Online or YouTube.

Downloading and analyzing output from a DISA developed SCAP scanner for the client or server operating system(s) on your home computer is also a value added suggestion to consider. It will enable you to understand how scanners measure the hardened configuration of a system while at the same time presenting output in the context of the NISP SP 800-53, Rev 4 requirements baseline.

It's also valuable to simply learn all the security relevant features and how to configure them for client or server operating systems you utilize in your job, at school or @ home. You could also partner with a trained system administrator, for a reasonable hourly rate, to learn how to enable/disable security relevant features of these Commercial Off The Shelf (COTS) operating systems and debate the risk mitigation value of each decision in context of their value in mitigating risks of the insider threat (on a small or very large corporate network).

You should also pro-actively take the initiative to periodically ask peers or review recently posted job announcements on the internet to determine the evolving portfolio of COTS or freeware security relevant tools employers expect their security staff to know how to utilize, day-one on the job.

"The value of a mentor in the Cybersecurity profession cannot be understated. However, the two critical factors to consider are 1) who is the best mentor given your flight time in the domain and 2) what strategy do you need to apply to ensure you get the highest value from the exchange.

Most mentors will be offended if it appears you simply want them to find you your next job or get you that promotion. They also will likely let you know quickly if they feel like you are observing them as they do most of the heavy lifting during your discussions."
– Carl L. Freeman, former Chief Information Security Officer (CISO); RAND Corporation

-13-

*"Computing forensics evidence collection can be a very tedious, boring task. In other words, you may spend multiple days or weeks, searching for that one small piece of forensic evidence you're looking for. You have to have patience and tenacity to succeed in this role. You'll also need to be willing to read the deeper level detail in vendor technical documentation for the unique platforms you'll be required to assess and the tools you'll be asked to apply in your job or you'll never be able to leverage their full capability."- **Fred Vickstrom, former Computing Forensics Investigator; The Boeing Company***

*"The largest risk is your Cybersecurity skills portfolio you'll lack if you allow yourself to primarily focus on forensics. Companies who have a large Cybersecurity budget can afford the luxury of staff with this unique skill set. In tough financial times, which inevitable come, investigators are typically the first to be laid off at medium to large firms because there is no mandatory requirement for their services"- **Carl L. Freeman, former Corporate Information Assurance Officer (CIAO); Booz Allen Hamilton***

What are some of the risks computing or network forensics investigations professionals must consistently focus on if they choose this career path?

Skill obsolescence is one risk that must be proactively mitigated. To reduce vulnerability, regularly take the time to review job announcements in this domain to learn about the evolving scope of tools utilized in computing or network forensics investigations. Network with other peers who utilize these tools and seek opportunities to get face-to-face training formally or in a mentor arrangement.

A very high percentage of the annual CyberSec budget is typically allocated to network forensics. Tools fielded to detect the external threat (the Chinese People's Liberation Army (PLA), for example). Little if any annual budget is allocated to increase effectiveness of computing forensics capabilities. Specifically, tools that enable detection of the insider threat and capture of this specific type of data for later presentation in a court of law.

If a CyberSec professional chooses the computing versus network forensics career path, they will likely have fewer opportunities for growth if they don't expand their skills beyond this domain and will also see fewer opportunities with other employers.

"Cybersecurity professionals should seriously consider sitting for a PMP exam and continuing to improve their project management skills as they progress through their career (https://www.pmi.org/certifications).

*All companies or agencies apply project management a little differently, so your ability to leverage their processes quickly will be critical to your success. Leadership thinks in project management / Agile terms, so the ability to create project management deliverables that enable effective project execution, quality and cost containment will enable you to be effective as you are asked to manage a wider scope of accountabilities in the future "- **Carl L. Freeman, former Corporate Information Assurance Officer (CIAO); Booz Allen Hamilton**

-14-

*"A Cybersecurity professional should have a library of hardcopy and e-books at their finger tips at home, at the office and on their company owned laptop or workstation. They should be reading a new book each week that contributes to solidifying skills they need at present or skills they know their employer will need them to leverage in the future as well. "- **Anonymous***

*"One great source to consider – Reading lists you can access online for undergraduate or graduate level courses in Cybersecurity offered at leading universities such as Stanford, USC, Harvard, George Washington University or MIT" – **Carl L. Freeman, former Systems Security Engineer (SSE) – French AWACS; Boeing Defense & Space Group***

What are the top three books on the market a Cybersecurity professional should have in their personal library?

Network Security Essentials – Applications and Standards (6th Edition) – William Stallings clearly presents under the hood detail for a wide range of security technologies to those who have deep technical depth and those with shallow hands on experience at the keyboard. It's a great reference if you need to review the basics of CyberSec tools in your architecture. This text book has been baked into the syllabus of several courses offered at multiple highly respected institutions that offer on-campus and online undergraduate and graduate degrees

'The Official (ISC)2 Guide to the CCSP CBK' (2nd Edition) – This high value reference is updated regularly by the most respected IT security certification authority, world-wide

All in One CISSP Exam Guide (8TH Edition) – Shon Harris's last release of this edition prior to her passing away in 2014 is a must for any Cybersecurity professionals library. She provides deep depth detail for a wide range of cybersecurity concepts and technologies. The research applied by the team who created the content is high value and is a great source for developing customized training artifacts for your Cybersecurity staff or employees at your company / agency.

Cybersecurity professionals should also consider books noted on the **Cybersecurity Canon** website as well. Book titles can be viewed at https://cybercanon.paloaltonetworks.com

Guide to Computer Forensics and Investigations: 6th Edition – This text is a great resource for a 'deep dive' into how data is stored in non-volatile memory and if required, how to capture it properly to support internal or external investigations sponsored by organization leadership or presentation as evidence in State / Federal courts

"Certificate programs offered by reputable institutions are a great way to refresh the education section of your resume with high value content. Carnegie Mellon and Stanford University are two examples of respected academic institutions that offer industry recognized Cybersecurity certificate programs that don't require a multi-year commitment to complete"- **Anonymous**

-15-

*"Most don't realize DoD and Intelligence Community A&A is a System Security Engineering (SSE) discipline at its core and it applies to weapons systems, satellites / ground stations, airborne and submersible platforms, software defined radios utilized by the Warfighter in the battlespace and the business systems / software development environments (SDEs) required to develop these deliverables to the U.S. Government"- **Anonymous**

"Assessment & Authorization is a unique skill set that can only be applied at a small number of defense contractor firms or a U.S. Government Intelligence Agency or a branch of the U.S. Department of Defense. Therefore, it can be more difficult to find opportunities if you are unwilling to relocate where the work is or you've damaged your reputation in the past" – **Carl L. Freeman, former Corporate Information Assurance Officer (CIAO); Booz Allen Hamilton**

What companies or Federal agencies hire CyberSec professionals with 'Assessment & Authorization' expertise, as defined by CNSSI 1253 and the U.S. Department of Defense?

- The Boeing Company
- Northrop Grumman
- The United States Navy
- Honeywell
- Booz Allen Hamilton
- RAND Corporation
- Defense Counterintelligence and Security Agency (DCSA)
- United States Army
- Amazon Web Services
- L3 Technologies
- Leidos

- Microsoft
- Lockheed Martin
- Textron
- SAIC
- General Dynamics
- Mandiant
- U.S. Marine Corps
- National Security Agency
- National Reconnaissance Office (NRO)
- United States Marine Corps
- U.S. Air Force
- Raytheon Technologies
- BAE

*"The fundamentals taught during U.S. Navy Boot Camp can serve as the foundation of personal behaviors that will result in a successful career in the Cybersecurity domain. For example, learning to consistently not be late, listening closely to the details when orders are delivered, realizing it's critical to be a member of a team focused on a group objective versus focusing on your own success are a few of many examples that will pay dividends for years to come"- **Anonymous**

60-Minute Career Path Deep Dive

CYBERSECURITY©

-16-

"Cybersecurity staff must rely on ITAR and/or Export Control training provided by subject matter experts in both domains. They should also seek concurrence from these experts before they implement technical or procedural controls intended to protect this sensitive data in all forms, including hard copy, in transit over the internet or data at rest on a CD or solid state drive"- **Jerry Beiter; former Director, Security & Export Compliance, Booz Allen Hamilton**

"Exposure of Sensitive but UCLASSIFIED data, governed by ITAR or U.S. Export Controls, is by default linked to and mitigated by application of 'Confidentiality' technical, personnel and physical controls addressed in the NIST SP 800-53r4 requirements baseline" - **Carl L. Freeman; President, (ISC)2 Orange County Chapter**

Should a Cybersecurity professional familiarize themselves with the fundamentals of the ITAR and Export Control regulations?

Familiarizing yourself has value because the more the more you know, the more you will be prepared to look deeper into these topics if you are suddenly accountable to know how to protect data of this type in all its forms, including e-form, hardcopy, visual/audio or integration into products/services provided to U.S Agencies, companies or international customers.

Companies commonly apply similar technical / procedural protections to ITAR/Export Controlled data that the U.S Government requires contractors to implement to protect 'Controlled UNCLASSIFIED Information (CUI)', as defined at the National Archives website, https://www.archives/cui.

It can be difficult to know the difference between CUI and information that must be protected in accordance with ITAR / Export Control regulations. This is why it's important to seek council from a trained professional in this domain.

Many top tier defense contractors have been fined or in extreme cases, had their authorization to compete for future U.S. Government contracts withdrawn for multiple years due to inconsistent application of protections or simply ignorance of how to apply the compliance criteria at the tactical level.

Several companies have assigned implementation of these regulations to the Law department simply because the financial impact of a violation can be devastating to a company unless they have the financial resources to absorb the financial and reputational impact of a violation.

-17-

*"Challenges and threats facing our country change almost daily. A key component of our national defense is well informed leadership which depends on accurate and concise information from the Intelligence Community (IC). Countless data bases are relied upon on to make informed decisions and this information is under constant attack. The challenge for today's Cybersecurity professional is effective 24x7x365 protection of this data from constantly evolving threats. The demand for competent Cybersecurity staff and leadership will only continue to increase as these threats continue to grow." - **Art Davis, former Director, Security; National Reconnaissance Office (NRO)***

*"Rewarding staff, middle, senior and executive level management opportunities, working with some of the most interesting look-forward technologies most in the Cybersecurity career domain will never see await in multiple U.S. Government Intelligence agencies for those willing to serve their country" – **Carl L. Freeman, former Data Processing Technician, 2nd Class (DP2 AW)); United States Navy***

Is Cybersecurity a viable, rewarding career path if one chooses a long term career with a U.S. Intelligence agency?

Yes. Several of the same Cybersecurity threats that apply at commercial entities such as Ford, Northrop Grumman, Microsoft, Boeing or Palo Alto Networks apply at U.S. Intelligence agencies as well.

The Chinese People's Liberation Army (PLA) and other state sponsored entities such as Iran/North Korea and Russia seek opportunity to exploit vulnerabilities in commercial corporate networks that enable sensitive or company proprietary data theft.

Since U.S. Intelligence agencies are obligated to protect U.S. National Security, the sensitive data they have in various forms is of particular value to nation states who do not support our values or democratic form of government.

A rewarding Cybersecurity career path at the staff or Selective Executive Service (SES) rank is available in domains such as encryption technology, Information Warfare, Cybersecurity policy/education & training or Systems Security Engineering (SSE) for a wide variety of ground based, air or space platforms.

The importance of an almost 'un-blemished' police record before and after employment with these U.S. agencies cannot be over-emphasized. Cybersecurity representatives who work for these organizations must accept that they will have to exercise due-diligence at all times to ensure they do not allow themselves to participate in behaviors that wouldn't be considered a risk at Twitter or Facebook.

*"If you don't intend to teach at the University level in this domain, you should seriously consider the multitude of personal sacrifices it will take to pursue a Cyber PhD. You should also determine if you're motivated by your intent to improve your value to a company or agency and its competitive viability or simply feed your ego"- **Ron Miller, former Information Technology Security Officer (ITSO); U.S. Navy – Puget Sound Naval Shipyard (PSNS)***

*"Higher level education not only opens up doors, it brings door handles you can see realistically within your reach" – **Carl L. Freeman, former Data Processing Technician, 2nd Class (DP2 AW)); United States Navy***

Is there value in pursuing an on-campus or online PhD in the Cybersecurity Domain?

Absolutely. There are multiple reasons why this is a great long term strategy, however, take care to select a school that doesn't have a reputation as a 'Diploma Mill':

- You could teach as a part-time career (or full time if you choose)
- Employers will recognize the value as you compete for upward advancement or new positions in other agencies or companies
- You can apply time spent to maintain your annual certification maintenance points for Information Technology (IT) security certifications (CISSP, CASP, etc.)
- You will expand your scope of skills. Specifically, you will likely leverage industry recognized Commercial Off the Shelf (COTS) security applications as you navigate through tasks defined in the syllabus and you'll learn valuable skills concerning how to document your research efforts
- You will be able to apply research skills you gain to your ongoing responsibilities with your employer

- You will differentiate yourself from 90+% of the competition that seeks the teaching or employment opportunities you apply for
- You will likely be approved to present topics of interest at industry recognized Cybersecurity conferences
- Staff who report to you will recognize your commitment to lifelong learning and will likely pursue additional education which will likely result in retention of staff in the organization your support
- Executive and senior leadership in your agency or company will likely assign you challenging tasks that have visibility enterprise wide

-19-

"Your integrity and police record will matter if you choose the Cybersecurity career path. So take care to keep your nose clean"
– Anonymous

"There are several positions in the U.S. Armed Forces, Cleared defense contractors like Northrop Grumman / Boeing you won't be able to apply for and if you do, you won't get an interview if you have a felony or in some cases certain misdemeanor convictions on your record. Take care to not close the door on this large quantity of job opportunities you could compete for in the future" " - **Carl L. Freeman, former Chief Security Architect – FIA Program, Ground Segment; Boeing Defense & Space Group**

Is my personal history relevant to beginning or maintaining momentum for a successful Cybersecurity career?

Cybersecurity is a career path that is heavily dependent upon a) credibility with employers/agencies due to consistent ethical behavior gained over time and b) a strong, unwavering commitment to ethical conduct on and off the job.

Eventually and often at the junior level, Cybersecurity staff will be asked to regularly interface with executive and senior level leadership in an agency or corporation. Credibility and a solid reputation will be critical to successful engagements with leadership

It's also critical to point out personal history can result in not being considered for a wide variety of Cybersecurity roles in a company/agency due to the sensitivity of information appointees must access regularly and commit to not share with those who don't have a legitimate, legally established need-to-know.

Some positions are contingent upon an employee's ongoing commitment to personal and professional conduct that does not result in the appearance of impropriety or a willingness to exercise bad judgment. A Driving Under the Influence (DUI) conviction after assuming a position or confirmation of lying on an employment application will likely result in a demotion and possible immediate termination.

The key takeaway to remember is: Personal and professional history matter at all times. Before you are hired and after you've assumed the role. Always build and protect your 'brand' and don't forget all action influence your overall reputation at work and beyond.

-20-

Can I pursue a Cybersecurity career without a 4 year college degree?

Yes, however, it isn't recommended. While many have succeeded with a 4 year degree, few have been promoted to middle management or the senior rank of CISO or Chief Security Officer (CSO) because they lack the pre-requisites to even apply for a position at the junior ranks and they typically lack a solid foundation in writing skills, business acumen and research expertise necessary to engage with senior technologists or executive / C-suite level leadership.

A degree in Physical Education or Political Science has value in this context, however, the immediate return on investment of a degree in Information Systems Management (ISM), Computer Science or Cyber-security cannot be understated.

Most who have completed their four year degree conclude quickly a Masters degree in a technical discipline such as Systems Engineering or Cybersecurity is critical if their undergraduate degree isn't in a technical domain or they need to compete with peers who have graduate level degrees and industry recognized Information Technology (IT) certifications.

Most who pursue a Cybersecurity career without a degree acknowledge, usually several years into their career, the many opportunities they couldn't apply for, the many times their manager or other senior leader pointed out they couldn't promote them or give them a significant salary increase to match their peers even though they could perform tasks at the same level of competency and quality.

"If you are a Cybersecurity professional or a Security Guard, always keep personal relationships out of the workplace. Never date a peer employee, someone you supervise, a client you support or someone who works at the same campus.

Many have taken this gamble and have regretted it. In some cases they have permanently damaged their reputation with their organization or those they must interface with / rely upon for their success in the organization for years to come."
— ***Anonymous***

60-Minute Career Path Deep Dive

CYBERSECURITY®

-21-

"If you are a junior level Cybersecurity analyst, ask your manager to teach you how the company manages their financial resources in the context of the organizations statement of work. Make this a habit. You'll be glad you did when you are suddenly accountable to serve as a Cost Account Manager"- **Anonymous**

"Cybersecurity analysts need to understand early in their career that their services must result in a return on investment / contribution to the business objectives of the organization they support. They should learn to evaluate their activities and other stakeholders who contribute to achievement of performance goals in context of the company or agency business management processes that contribute to achievement of the overall mission" – ***Carl L. Freeman, former Chief Information Security Officer (CISO); RAND Corporation***

Are business skills critical for a successful CyberSec career?

At the junior/mid-level Cybersecurity analyst level business skills will not be required because managers will typically engage with senior / executive organization leadership to project and manage the 'cost actuals' linked to an annual CISO authorized budget.

Mid-level Cybersecurity analysts should ask their managers to engage in budget management activities so they can observe how the 'sausage is made', real-time. If the analyst applies for a management position, they can highlight their efforts to collaborate with management to understand business management processes, tools and performance metrics.

Every company has its own business management processes and they are rarely designed to be user friendly for Cybersecurity management. This is why it's advisable to assertively pursue all opportunities to learn the business processes that apply at the agency or company you support.

This skill can be of very high value if the organization does not adequately staff the organization to meet the full scope of product/service and more importantly service level expectations. When this scenario applies, the Cybersecurity lead or manager needs to be able to rely on performance and business metrics collected over time if they hope to justify an increase in budget or labor resources.

-22-

Are Lean / Agile principles and their tactical implementation in context of CyberSec important to understand?

Lean / Agile has been around for a long time. However, it's accelerating relevance in the U.S. government, state / local governments and contractors who provide services to both justifies a proactive strategy to not only learn the principles, but how they are being leveraged in the environments Cybersecurity professional's support.

Cybersecurity staff and the managers they support traditionally don't see value in initiatives such as Lean / Agile or other 'Fad' programs senior leadership believes have value every few years. This is a critical mistake that all should beware of and if it's detected, it needs to be addressed immediately. Computers and software and network devices are 'sexy', however, Cybersecurity staff need to understand the organization exists to make a profit or provide a product/service. The organization doesn't exist to simply enable us to 'Play with the Latest Toys' on the market.

As noted in the same context for 'business management' processes, most organizations have a unique application of Lean/Agile that could be linked to an industry recognized framework like SAFe 5.0, for example.

Most Cybersecurity professionals with 15+ years of experience struggle with grasping the tactical level implementation of Lean / Agile in the context of their responsibilities to their customers and how their individual or organizations success is measured. The natural tendency to experience stress when engaging in 'change' needs to be acknowledged with the understanding this will change over time as all stakeholders learn the relevant tools, language and value of Lean / Agile.

23

*"Some are surprised how critical writing skills are in the Cybersecurity domain. Working at the keyboard, leveraging Cybersecurity COTS tools in a production environment is only critical for a small number of possible career paths in the Cyber domain. If one moves into first level, senior or executive Cyber leadership, technical skill is of moderate importance"- **Anonymous***

"IT Security policy, guidance or training artifacts posted on public domain websites represents the CISOs attempt to communicate the applied security requirements baseline most would not understand or find as exciting as watching paint dry to internal stakeholders accountable to comply and the customers the organization serves as well."- **Carl L. Freeman; President, (ISC)[2] Orange County Chapter**

Why is it valuable to review CyberSec policies posted on U.S. Federal Government, U.S Military, State or Public/Private University websites?

Most of these organizations have a legal obligation to protect sensitive types of information, such as Personal Health Information (PHI), Personal Financial Information (PHI) or Controlled UNCLASSIFIED Information (CUI).

Senior Information Security executive leaders are accountable for not only releasing this type of policy or guidance to their staff and internal/external customers, they are also required to ensure it aligns with industry best practices or legislative imperatives such is ISO 270000 or NISP SP 800-53, Revision 4 or 5.

Analysis of policy / guidance posted on public domain websites maintained by these entities is of particular value when you consider the relevant technical control baseline authors likely applied to create the policy or security training material which may also be posted for all stakeholders to see on their .gov, .edu or .mil websites.

If you want to learn Cybersecurity terminology, discover how the different security requirements baselines and state/Federal Cyber law are implemented at the tactical level, what better way to see how the 'sausage is made' then to review policies, standards and procedures organizations apply to protect the many different types of sensitive data they are obligated to protect.

24

*"Privacy is equal to Confidentiality in the Confidentiality-Integrity-Availability (CIA) foundation of Cybersecurity"- **Anonymous***

*"IT Security and Privacy program compliance programs must be joined at the hip to be successful. Privacy Act related information in hardcopy or e-form is simply a different flavor of sensitive data that requires protection. The only difference between the two is when there is a confirmed compromise of this data, stockholders will hold the CISO and CEO accountable if proprietary data is compromised. The State Attorney General or Federal Government agency information owner (IO) will hold these two executives accountable if the data compromised is employee or customer PHI/PII"- **Carl L. Freeman, former Chief Information Security Officer (CISO); RAND Corporation***

Are there Cybersecurity career opportunities if you are a Privacy professional?

Absolutely. There are several common accountabilities shared by the appointed CISO and Chief Privacy Officer of an organization. Implementation of technical and/or procedural controls to protect information that meets the criteria of U.S. or State 'Privacy Act' related legislation are often assigned to the CISO by default in most smaller organizations.

Some firms assign management of the privacy program to the Office of Legal Council and obligate the Chief Legal Officer to proactively collaborate with the CISO to define where the data is stored/transmitted, what technical/procedural controls are applied to protect it in hardcopy or e-form and what steps will be taken on a recurring basis to determine success of the Privacy compliance efforts.

CISOs and Privacy Officers are also accountable to document and test procedures that align with internal or Federal / State mandated 'Incident Response' accountabilities. These procedures define how to respond effectively if there is a Personal Health / Personal Financial Information (PHI/PFI) data breach triggered by an Insider Threat in the organization or an external entity (hacker or member of a nation state sponsored organization like the Chinese People's Liberation Army (PLA)).

-25-

"Reserve time on your calendar weekly to document a) what you need to know to do your job well and b) don't know at present and execute a self-directed training plan that effectively addresses these ongoing deficiencies that will change on a regularly basis."- **Anonymous**

"If you don't like to study, take classes online or instructor-led on a regular basis and teach others what you know, you shouldn't pursue Cybersecurity as a career path." — **Carl L. Freeman, former Chief Security Architect – FIA Program, Ground Segment; Boeing Defense & Space Group**

Why is a commitment to Life-long learning critical if one chooses to pursue a career in Cybersecurity?

Companies who sell technology to companies, State / Federal agencies and citizens must improve their products and add additional features or they will not remain competitive in the market. As companies develop enhancements or significant improvements to their products, the security vulnerabilities to data created, stored, processed or transmitted on these 'products' changes.

Cybersecurity professionals will have an ongoing commitment to master a given security related product, like SPLUNK, or upgrade their current portfolio of capabilities/skills they can provide to their company or customers.

Some don't take this ongoing commitment seriously and find over time they don't get job offers when they interview for positions and they aren't offered upward mobility in the organizations they support.

It's important to note life-long learning in the Cybersecurity domain isn't limited to technology. It also includes learning or refining soft skills like how to be a valued member of a team, how to present to executives in your organization or how to contribute effectively in a Lean/Agile environment.

Senior leadership will compare your actions concerning maintenance of your skill set as they consider your future assignments and upward mobility. If an employee doesn't pursue ongoing skill enhancement and knowledge of leading edge security tools and the evolving security threat, they eventually contribute to the risk the Cybersecurity organization is accountable to mitigate.